ANIMALS OF THE TUNDRA

FANTASTIC ARCTIC FOXES

THERESA EMMINIZER

PowerKiDS press.
New York

Published in 2022 by The Rosen Publishing Group, Inc.
29 East 21st Street, New York, NY 10010

First Edition

Portions of this work were originally authored by Maeve T. Sisk and published as *Arctic Foxes*. All new material in this edition authored by Theresa Emminizer.

Editor: Jill Keppeler
Book Design: Michael Flynn

Photo Credits: Cover, p. 1 Schaef1/iStock/Getty Images; (series background) Stefan Sorean/Shutterstock.com; p. 5 JoannaPerchaluk/Shutterstock.com; p. 7 Troutnut/Shutterstock.com; p. 9 DmitryND/E+/Getty Images; p. 11 (main) pum_eva/iStock/Getty Images; p. 11 (inset) Robert Postma/Design Pics/Getty Images; p. 13 Trond Eriksen/500px/Getty Images; p. 15 Danita Delimont/Gallo Images/Getty Images; p. 17 Caroline Evensen/EyeEm/Getty Images; p. 19 Giedriius/Shutterstock.com; p. 21 Warren Metcalf/Getty Images.

Library of Congress Cataloging-in-Publication Data

Names: Emminizer, Theresa, author.
Title: Fantastic arctic foxes / Theresa Emminizer.
Description: New York : PowerKids Press, [2022] | Series: Animals of the tundra | Includes index.
Identifiers: LCCN 2020021714 | ISBN 9781725326279 (library binding) | ISBN 9781725326255 (paperback) | ISBN 9781725326262 (6 pack)
Subjects: LCSH: Arctic fox–Juvenile literature. | Tundra animals–Juvenile literature.
Classification: LCC QL737.C22 E595 2022 | DDC 599.776/4–dc23
LC record available at https://lccn.loc.gov/2020021714

Manufactured in the United States of America

CONTENTS

Meet the Arctic Fox!

Arctic foxes are small and fluffy. They have white fur, short ears, and a thick tail. An Arctic fox's little body is specially made to hold in warmth. That's because it lives in a very cold **environment**—the tundra.

Life in the Tundra

The tundra is cold and windy. The **temperature** outside is usually -30° Fahrenheit to 20° Fahrenheit (-34° Celsius to -6° Celsius). There aren't many plants because the ground is always frozen. How are Arctic foxes able to live there?

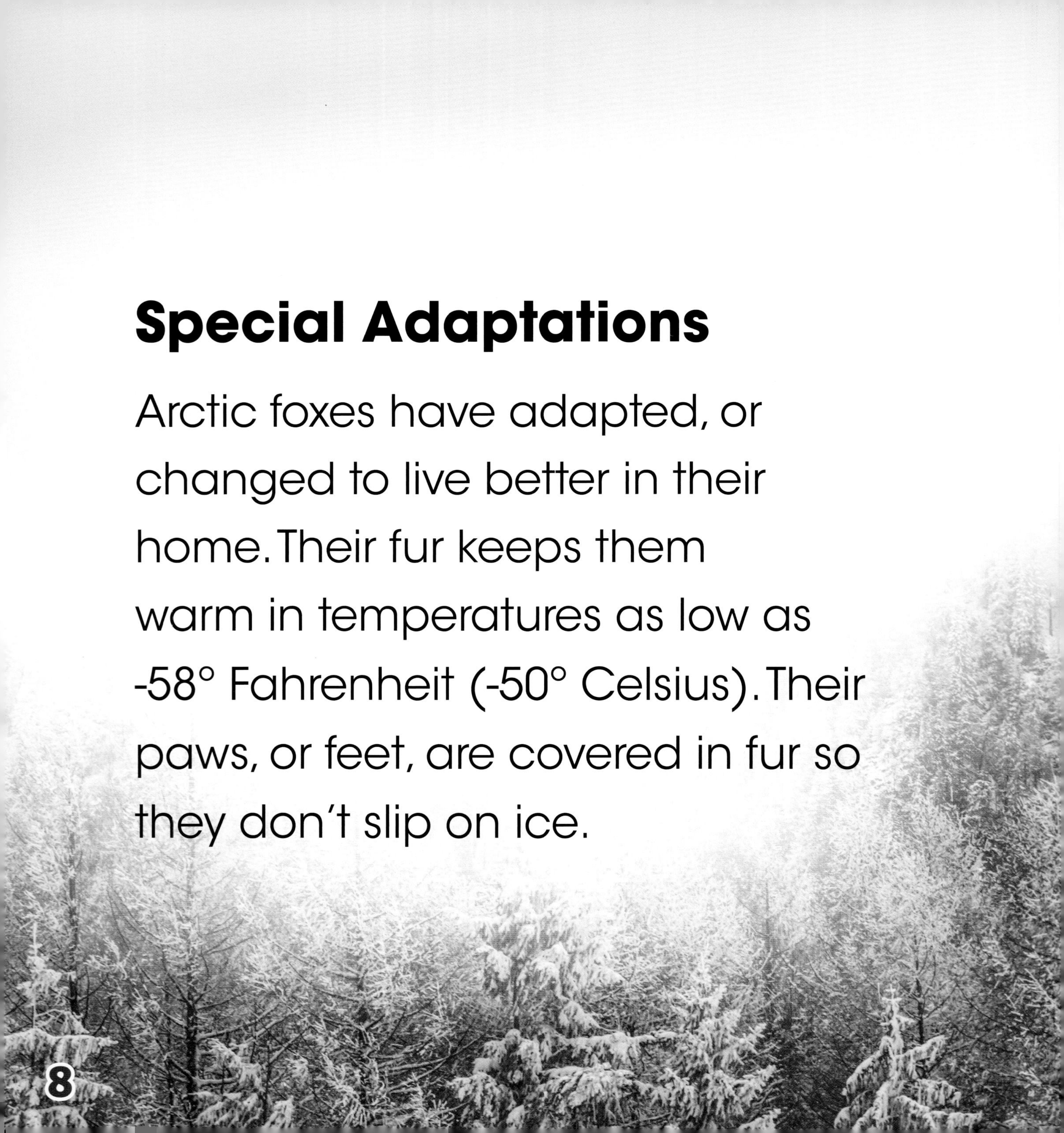

Special Adaptations

Arctic foxes have adapted, or changed to live better in their home. Their fur keeps them warm in temperatures as low as -58° Fahrenheit (-50° Celsius). Their paws, or feet, are covered in fur so they don't slip on ice.

Cool Camouflage

One adaptation is **camouflage**. Arctic foxes' fur changes color over the seasons. During winter, their fur is white to **blend** in with the snow. When the weather warms, their fur turns brown or gray to blend in with plants and rocks.

On the Hunt

Camouflage helps Arctic foxes stay safe from **predators**. It also helps them hide while they hunt. Arctic foxes mostly eat mouselike animals called lemmings. They also eat birds and fish. Sometimes they follow polar bears to eat their leftovers!

Arctic foxes' round ears help them hear lemmings moving underneath several inches of thick snow. They're also helped by their strong sense of smell. When they hear lemmings moving, Arctic foxes **pounce**, diving nose-first into the snow.

Fox Families

A group of Arctic foxes is called a skulk. Female, or girl, foxes are called vixens. Males, or boys, are called dogs. In springtime, males and females come together to mate, or make babies. Arctic fox babies are called kits.

A mother fox gives birth to 12 to 14 kits. Parents care for kits for the first few months. In the beginning, kits drink milk from their mother's body. Fox families live together in a den or burrow.

Fox Dens

Arctic fox dens are made up of tunnels and rooms dug under the soil. The foxes have many different entrances, or ways to get in. Entrances may be hidden under plants or rocks. Some dens are home to **generations** of foxes.

Fancy Features

GLOSSARY

blend: To look like what is around you.

camouflage: Colors or shapes on animals that allow them to blend in with their surroundings.

environment: The conditions that surround a living thing and affect the way it lives.

generation: The people or animals all born and living at about the same time.

pounce: To jump suddenly toward or onto something.

predator: An animal that eats other animals.

temperature: How warm or cold something is.

FOR MORE INFORMATION

WEBSITES

Arctic Fox
www.nationalgeographic.com/animals/mammals/a/arctic-fox/
Learn more about the Arctic fox and see color photos on this National Geographic website.

Arctic Fox: All in the Family
kids.sandiegozoo.org/animals/arctic-fox
Read more fascinating facts about the Arctic fox on this San Diego Zoo Kids website.

BOOKS

Pettiford, Rebecca. *Arctic Foxes*. Minneapolis, MN: Bellwether Media, 2019.

Pfeffer, Wendy. *The Artic Fox's Journey*. New York, NY: Harper Collins, 2019.

INDEX